Grade 1 Math Practice Counting and Comparing

1. Which symbol makes the correct comparison?

    6 _____ 66

    a. x
    b. =
    c. <
    d. >

2. Which of the following numbers is the least?

    a. 12
    b. 15
    c. 10
    d. 21

3. Compare: 4 _____ 14

    a. >
    b. <

4. Compare: 57 _____ 55

    a. <
    b. >

5. 7 < 10

    a. True
    b. False

6. 21 < 24

    a. True
    b. False

7. Write the number that comes just before.

    _____, 33

8. Which number comes right after?

    101, _____

    a. 90
    b. 99
    c. 100
    d. 102

9. 37 < ?
    a. 37
    b. 28
    c. 40
    d. 36

10. Ashely had 16 baseball cards, Johnny had 15 cards, Angel had 20 cards, and Martin had 34 cards. Who had the most baseball cards?
    a. Ashely
    b. Martin
    c. Johnny
    d. Angel

11. Which symbol correctly compares the numbers?

    45 _____ 54
    a. =
    b. <
    c. >

12. Which symbol correctly compares the numbers?

    76 _____ 76
    a. =
    b. <
    c. >

13. Compare.

    72 < 81
    a. True
    b. False

14. Which number is greater than 17?
    a. 11
    b. 21
    c. 12

15. Which number is less than 50?
    a. 60
    b. 40
    c. 80

16. Which one means greater than?
    a. >
    b. <
    c. =

17. 19_____11

18. How many mittens are there? Count by twos.

    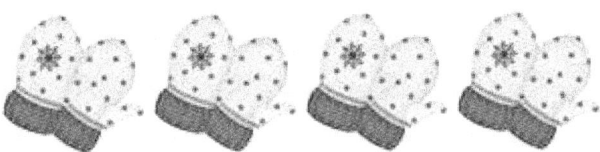

    a. 2
    b. 4
    c. 6
    d. 8

19. Which sign correctly compares the number of pictures?

    a. >
    b. <
    c. =

20.

    a. True
    b. False

21. 0_____ 9

    a. <
    b. >
    c. =

22. 86_____92

    a. <
    b. >
    c. =

23.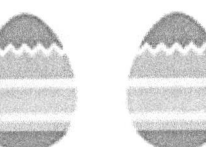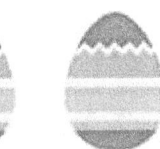

    a. True
    b. False

24. What does greatest mean?
    a. smallest
    b. biggest
    c. equal

25. Which number comes next?

    108, 109, _____
    a. 106
    b. 107
    c. 110
    d. 111

Grade 1 Math Practice Patterns

1. Write the numbers to complete the pattern. 42, 52, _____, _____, 82, _____

2. I am counting by 2s. What number fills the blank? 2, 4, 6, 8, 10, __, 14

    a. 13
    b. 21
    c. 8
    d. 12

3. Skip count to find the number that comes next.

    88, 90, 92, _____

4. Bryan is counting by 3s. He says the numbers:

    3, 6, 9, 12

    Which number should Bryan say next?

    a. 13
    b. 14
    c. 15
    d. 16

5. When skip-counting by 2, follow the pattern below.

    2, 4, 6, 8, _____

    What is the missing number?

    a. 7
    b. 12
    c. 10
    d. 20

6. Skip count by 2. Fill in the missing blanks.

    2, 4, _____, 8, _____, 12, 14, 16 _____, 20

7. You are helping your mom unpack groceries. Which food belongs in this group?

a.

b.

c.

8. What number comes next?

   57, 59, 61, ?

   a. 65
   b. 63
   c. 67

9. 2, 4, _____, 8

10. Choose the numbers that complete the pattern. 77, 67, 57, _____, _____, _____

    a. 47, 37, 27
    b. 56, 55, 54
    c. 87, 97, 107
    d. 50, 40, 30

Grade 1 Math Practice Place Value

1. Which numbers round to 800?
    a. 100
    b. 799
    c. 804
    d. 710

2. Please fill in the blanks by identifying how many ones and tens are in the following number.

    93

    _____ ones

    _____ tens

3. Count the groups of tens and ones.

    How many tens? _____ How many ones? _____ How many blocks? _____

4. Circle a group of ten baseballs.

    How many tens? _____ How many ones? _____ How many baseballs? _____

5. Write the number.

   a) One ten = _____

   b) Nine tens = _____

6. Write the number of tens and ones.

   a) 75 = _____ ten(s) and _____ one(s)

   b) 20 = _____ ten(s) and _____ one(s)

   c) 99 = _____ ten(s) and _____ one(s)

   d) 18 = _____ ten(s) and _____ one(s)

   e) 10 = _____ ten(s) and _____ one(s)

7. Write the symbol >, =, or < on the line which correctly compares the sets of blocks.

a)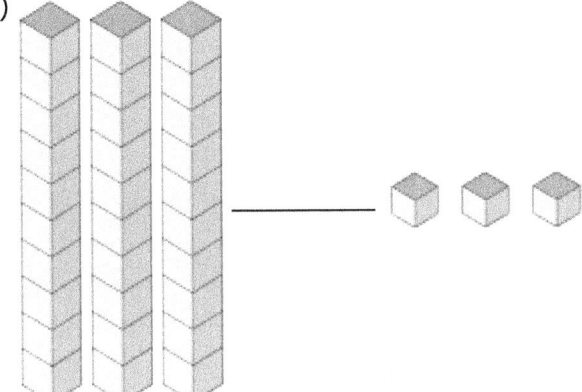

b)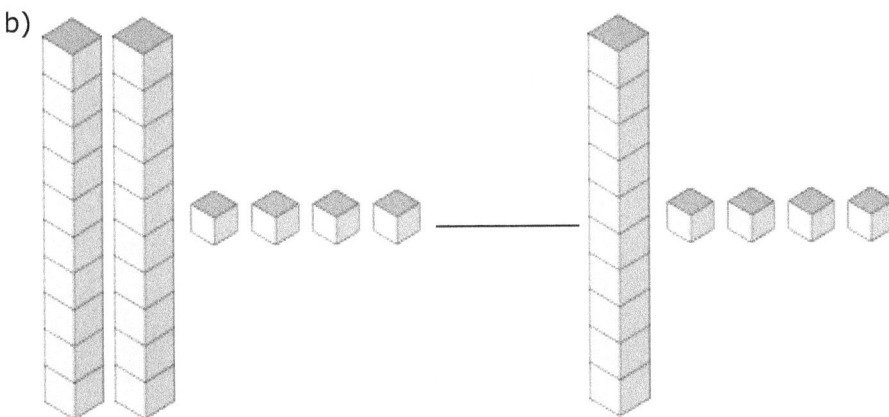

c)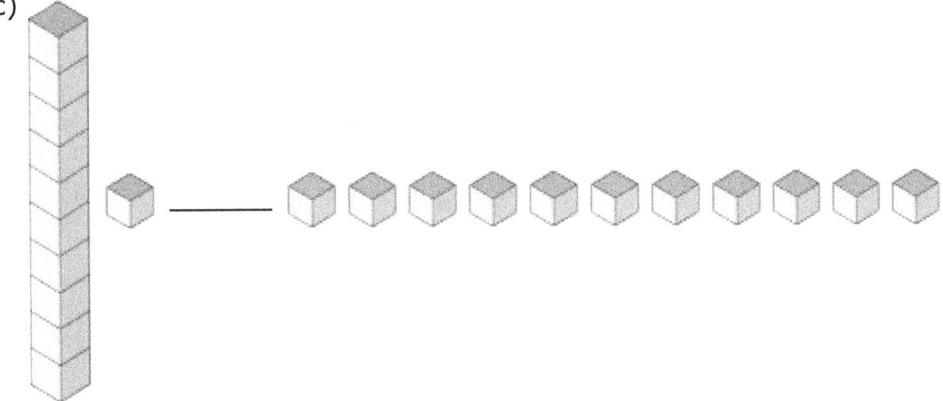

8. How many tens?

   a. 1
   b. 2
   c. 10
   d. 20

9. How many tens?

   a. 7
   b. 8
   c. 9
   d. 10

10. Count the slices of pie. Fill in the blanks.

   _____ tens + _____ ones = _____ slices of pie

11. Count the objects. Fill in the blanks.

_____ ten + _____ ones = _____ total

12. 51 has _____ tens and _____ one.

13. Lee collects marbles. He can make 11 groups of 10 marbles and has 2 marbles left over. How many marbles does Lee have?
    _____ marbles

14. How many tens are in the number fifty?

    a. 0
    b. 5
    c. 10

15. Which shows how to compare the numbers 10 and 13?

    a. $10 = 13$

    b. $10 > 13$

    c. $10 < 13$

16. Which shows 20?

a.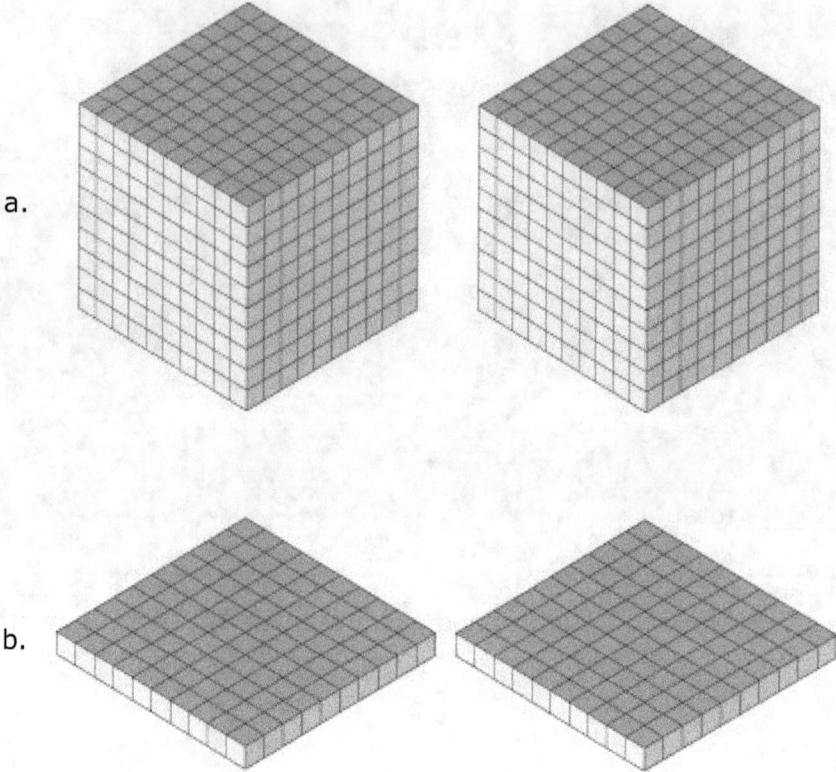

b.

c.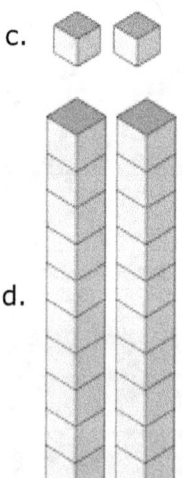

d.

17. Which shows 12?

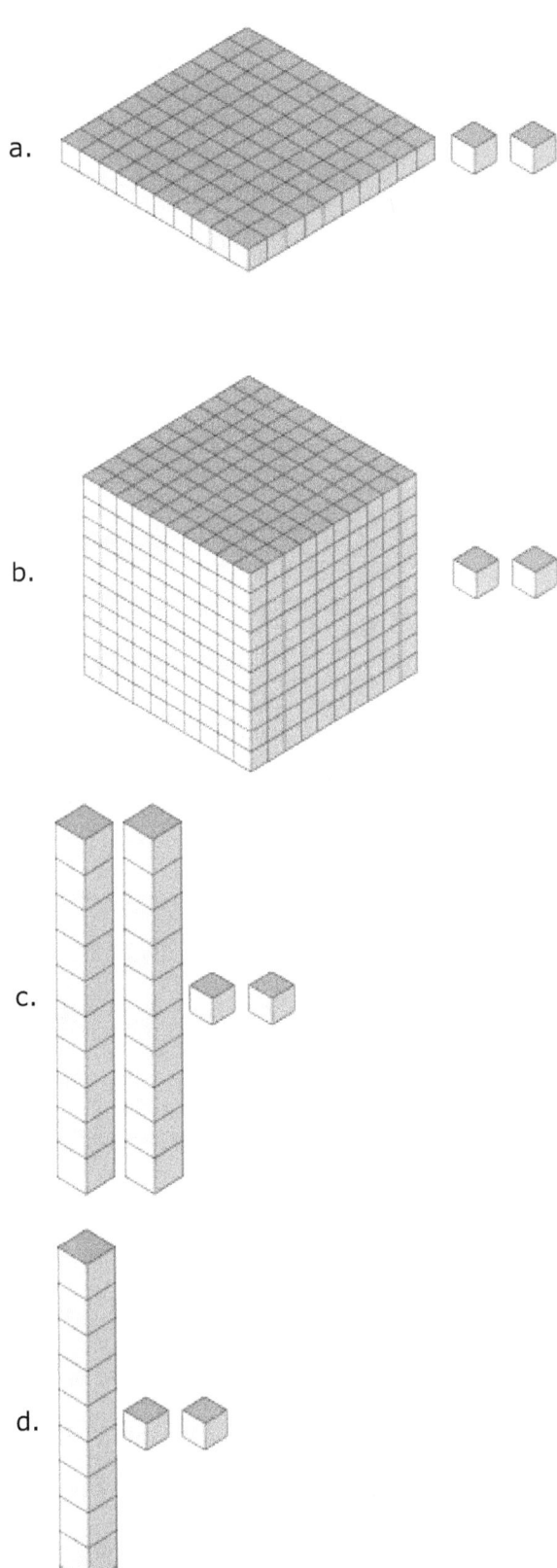

18. How many tens are shown?

   a. 4
   b. 5
   c. 1
   d. 2

19. How many is two tens and 5 ones?

20. Which numbers round to 70?
   a. 66
   b. 77
   c. 85
   d. 72

21.

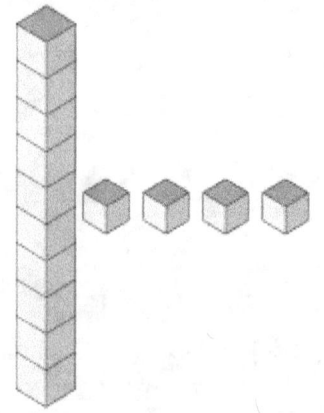

   How many tens? _____

   How many ones? _____

   How many cubes? _____

22.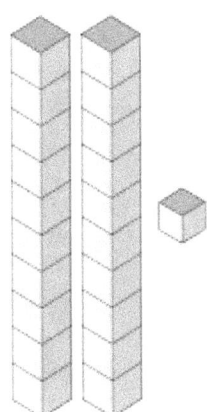

How many tens? _____

How many ones? _____

How many cubes? _____

23. Kai built the tower of blocks shown. Melanie knocked down the tower. Draw what the same number of blocks looks like now.

24. Write the number of tens and ones for each number.

   a) 19 = _____ ten(s) and _____ one(s)

   b) 11 = _____ ten(s) and _____ one(s)

25. How many tens?

   a) 40 = _____ tens

   b) 70 = _____ tens

Grade 1 Math Practice Fact Families

1. Which is the related addition fact for $4 + 8 = 12$?

    a. $6 + 6 = 12$

    b. $7 + 5 = 12$

    c. $8 + 4 = 12$

    d. $9 + 3 = 12$

2. Which is the related addition fact for $3 + 8 = 11$?

    a. $8 + 3 = 11$

    b. $5 + 6 = 11$

    c. $9 + 2 = 11$

    d. $6 + 5 = 11$

3. Which is the related addition fact for $5 + 4 = 9$?

    a. $1 + 8 = 9$

    b. $2 + 7 = 9$

    c. $3 + 6 = 9$

    d. $4 + 5 = 9$

4. Which is the related subtraction fact for the number sentence?

    $16 - 9 = 7$

    a. $9 - 7 = 2$

    b. $16 - 7 = 9$

5. Fill in the missing number.

    If 5 + 3 = 8,

    then 8 - _____ = 5

6.

$8 +$  $= 10$

$10 - 8 =$

 $= ?$

   a. 2
   b. 18

7. What is the related subtraction fact for $9 - 2 = 7$?

   a. $7 - 2 = 5$

   b. $8 - 2 = 6$

   c. $9 - 7 = 2$

   d. $10 - 3 = 7$

8. Which is the missing fact from the fact family?

   $3 + 7 = 10$
   $7 + 3 = 10$
   $10 - 7 = 3$

   a. $7 - 3 = 4$

   b. $10 - 3 = 7$

   c. $3 + 10 = 13$

   d. $10 + 7 = 17$

9. Which is the missing fact from the fact family?

   $12 - 4 = 8$
   $12 - 8 = 4$
   $4 + 8 = 12$

   a. $8 - 4 = 4$

   b. $8 + 4 = 12$

   c. $12 + 4 = 16$

   d. $12 + 8 = 20$

10. Which is the related subtraction fact for the number sentence?

    $15 - 11 = 4$

    a. $11 - 4 = 7$

    b. $15 - 4 = 11$

11.

   $5 +$ 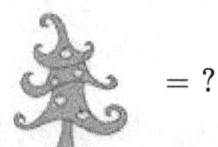 $= 5$

   $5 - 5 =$ 🎄

   🎄 $= ?$

   a. 5
   b. 0

12.

6 +  = 7

7 − 6 =

 = ?

   a. 13
   b. 1

13.

4 +  = 8

8 − 4 =

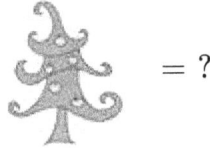 = ?

   a. 4
   b. 8

14. Fill in the missing number.

If 7 + 6 = 13,

then 13 - 6 = _____

15.

$4 +$  $= 7$

$7 - 4 =$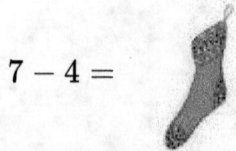

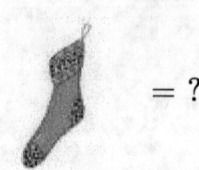 $= ?$

a. 11
b. 3

Grade 1 Math Practice Fact Families Answer Key

1. Which is the related addition fact for $4 + 8 = 12$?

    a. $6 + 6 = 12$

    b. $7 + 5 = 12$

    c. $8 + 4 = 12$

    d. $9 + 3 = 12$

2. Which is the related addition fact for $3 + 8 = 11$?

    a. $8 + 3 = 11$

    b. $5 + 6 = 11$

    c. $9 + 2 = 11$

    d. $6 + 5 = 11$

3. Which is the related addition fact for $5 + 4 = 9$?

    a. $1 + 8 = 9$

    b. $2 + 7 = 9$

    c. $3 + 6 = 9$

    d. $4 + 5 = 9$

4. Which is the related subtraction fact for the number sentence?

    $16 - 9 = 7$

    a. $9 - 7 = 2$

    b. $16 - 7 = 9$

5. Fill in the missing number.

    If 5 + 3 = 8,

    then 8 - _____ = 5

    3

6.

$8 +$  $= 10$

$10 - 8 =$

 $= ?$

   a. 2
   b. 18

7. What is the related subtraction fact for $9 - 2 = 7$?

   a. $7 - 2 = 5$

   b. $8 - 2 = 6$

   c. $9 - 7 = 2$

   d. $10 - 3 = 7$

8. Which is the missing fact from the fact family?

   $3 + 7 = 10$
   $7 + 3 = 10$
   $10 - 7 = 3$

   a. $7 - 3 = 4$

   b. $10 - 3 = 7$

   c. $3 + 10 = 13$

   d. $10 + 7 = 17$

9. Which is the missing fact from the fact family?

   $12 - 4 = 8$
   $12 - 8 = 4$
   $4 + 8 = 12$

   a. $8 - 4 = 4$

   b. $8 + 4 = 12$

   c. $12 + 4 = 16$

   d. $12 + 8 = 20$

10. Which is the related subtraction fact for the number sentence?

    $15 - 11 = 4$

    a. $11 - 4 = 7$

    b. $15 - 4 = 11$

11.

    $5 +$ 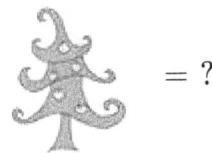 $= 5$

    $5 - 5 =$

     $= ?$

    a. 5
    b. 0

12.

6 +  = 7

7 − 6 =

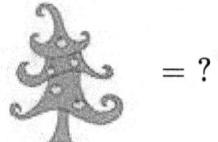 = ?

a. 13
b. 1

13.

4 + ![tree] = 8

8 − 4 = ![tree]

![tree] = ?

a. 4
b. 8

14. Fill in the missing number.

If 7 + 6 = 13,

then 13 - 6 = _____

7

15.

4 +  = 7

7 − 4 =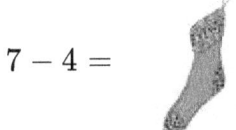

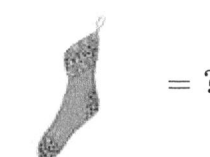 = ?

a. 11
b. 3

Grade 1 Math Practice Money Counting

1. I am worth five cents.
    a. Penny
    b. Nickel
    c. Quarter
    d. Dime

2. I am worth ten cents.
    a. Penny
    b. Nickel
    c. Quarter
    d. Dime

3. Which amount is the greatest?
    a. $34
    b. $43
    c. $40
    d. $30

4. If each coin shown is worth 5 cents, how many more coins would you need to show 30 cents altogether?

    a. 0
    b. 5
    c. 3
    d. 1

5. How much money is shown?

    a. 21 cents
    b. 12 cents
    c. 3 cents
    d. 16 cents

6. I am worth one cent.
    a. Penny
    b. Nickel
    c. Quarter
    d. Dime

7. If each coin is worth 5 cents, how many cents are shown?

   a. 55 cents
   b. 65 cents
   c. 60 cents
   d. 70 cents

8.

   a. 20 cents
   b. 5 cents
   c. 25 cents

9.

   a. 65 cents
   b. 75 cents
   c. 35 cents

10. I am worth twenty-five cents.

    a. Penny
    b. Nickel
    c. Quarter
    d. Dime

11. Skip count to find the total value.

    a. 50 cents
    b. 25 cents
    c. 5 cents
    d. 20 cents

12. What is the value of this coin?

   a. 25 cents
   b. 5 cents
   c. 10 cents
   d. 1 cent

13.

   a. 20 cents
   b. 22 cents
   c. 12 cents

Grade 1 Math Practice Arithmetic & Number

1. Peter bought 7 apples. He ate 2 apples. He gave 1 apple to a friend. How many apples does Peter have left?

2. $12 - 3$

3. $9 - 3$

4. $10 + 4 =$
   a. 5
   b. 14
   c. 11
   d. 7

5. $15 + 1 =$
   a. 14
   b. 16
   c. 15
   d. 10

6. $20 + 4 =$
   a. 22
   b. 19
   c. 24
   d. 10

7. Mom cut the apple pie into 8 pieces. I ate 3 pieces and Sara ate 2 pieces. How many pieces of pie are left?

8. The Bulldogs scored 14 points during Thursday's game and 5 points on Saturday. How many more points did they score on Thursday compared to Saturday?

9. Amy received 2 stickers (each day) from her teacher on Monday, Wednesday, and Friday. She only received 1 sticker (each day) on Tuesday and Thursday. How many stickers did Amy earn for the week?

10. $13 - 6 =$

    a. 7
    b. 8
    c. 5

11. 3 _ 4 = 7

    a. $+$
    b. $-$
    c. $=$

12. Which symbol makes the correct comparison?

    6 _____ 66

    a. x
    b. =
    c. <
    d. >

13. Which numbers round to 800?

    a. 100
    b. 799
    c. 804
    d. 710

14. Which of the following numbers is the least?

    a. 12
    b. 15
    c. 10
    d. 21

15. Compare: 4 _____ 14

    a. >
    b. <

16. Compare: 57 _____ 55

    a. <
    b. >

17. 7 < 10
    a. True
    b. False

18. 21 < 24
    a. True
    b. False

19. Sophia has 14 lollipops. She gives 3 to Makenzi and 3 to Mari. How many lollipops does she have left?
    a. 8 lollipops
    b. 3 lollipops
    c. 10 lollipops
    d. 5 lollipops

20. Write the number that comes just before.

    _____, 33

21. $9 + 3 =$
    a. 12
    b. 6
    c. 27
    d. 11

22. $8 + 4 =$
    a. 4
    b. 32
    c. 11
    d. 12

23. $6 + 5 =$
    a. 1
    b. 30
    c. 10
    d. 11

24. $9 + 5 =$
    a. 4
    b. 45
    c. 14
    d. 15

25. $10 + 5 =$
    a. 5
    b. 50
    c. 2
    d. 15

Grade 1 Math Practice Geometry & Measurement

1. What time does the clock show?

2. What time does the clock show?

3. There are 90 minutes in one hour.
    a. True
    b. False

4. The short hand on the clock is called the _____ hand.
    a. hour
    b. minute
    c. second
    d. day

5. What time is shown on the clock?

   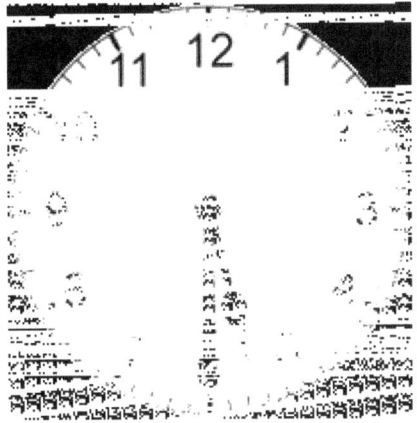

   a. 6:05
   b. 6:25
   c. 5:30
   d. 5:06

6. What time is shown on the clock?

   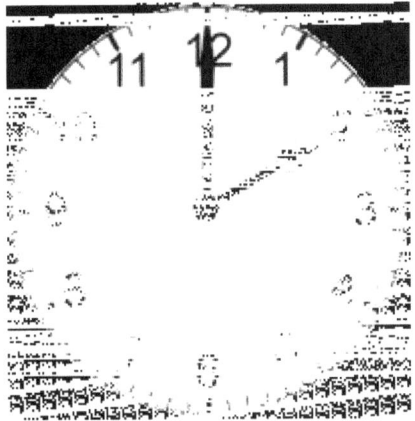

   a. 12:02
   b. 12:10
   c. 2:00
   d. 12:00

7. Which month comes right after March?

   a. March
   b. February
   c. April

8. How many corners does this shape have?

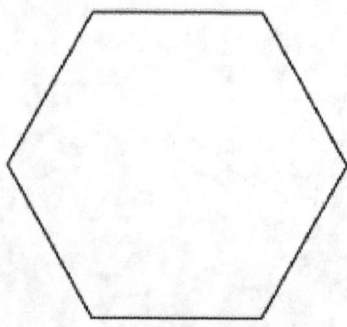

   a. 3
   b. 4
   c. 6

9. Identify the polygons that make-up the triangular prism shown. How many of each polygon make up the figure?

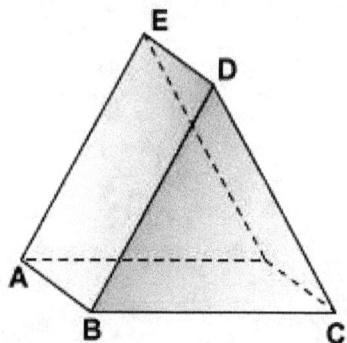

10. What is each division of time for the big hand on the clock?

11. What shape is this? o
    a. triangle
    b. square
    c. circle
    d. rectangle

12. Abby's room is 10 feet long. Emilio's room is 2 feet longer than Abby's room. Mark's room is 1 foot shorter than Abby's room. Who has the longest room?
    a. Abby
    b. Emilio
    c. Mark

13. The puppy puzzle is 18 inches long. The rainbow puzzle is 3 inches shorter than the puppy puzzle. The train puzzle is 2 inches shorter than the puppy puzzle. Which puzzle is the longest?
    a. puppy
    b. rainbow
    c. train

14. Draw the hands on the clock to show 3:00.

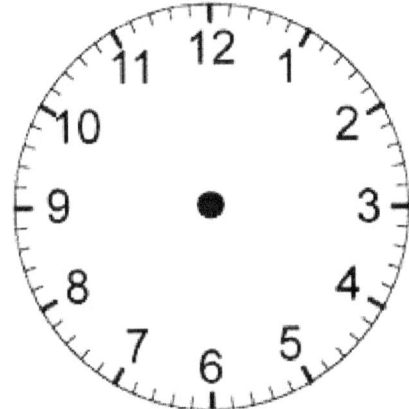

15. Draw a line on the figure that divides it into 2 trapezoids.

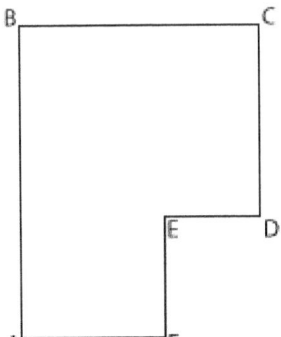

16. Draw a line on the shape to create 2 half-circles.

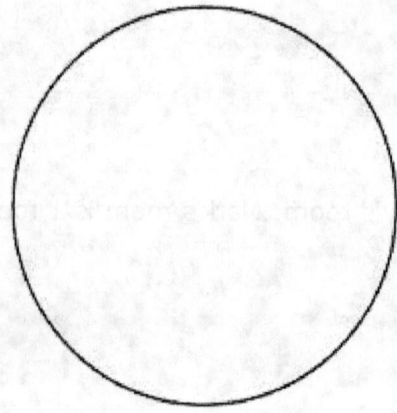

17. If you draw a line from A to E, what two shapes would you make?

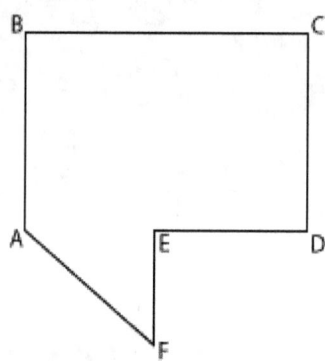

   a. trapezoid and square
   b. rectangle and trapezoid
   c. square and triangle
   d. triangle and rectangle

18. Look at the pencils:

   A.

   B.

   C.

   Which correctly orders the pencils shortest to longest?
   a. A, B, C
   b. B, C, A
   c. C, A, B

19. Circle the clock that shows 5 o'clock.

20. Circle the clock that shows 6 o'clock.

21. Which of the following would change the shape of the rectangle?

   a. color it blue
   b. make it smaller
   c. remove one side
   d. turn it clockwise

22. Which figure shows halves?

a.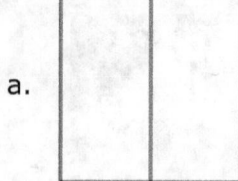

b.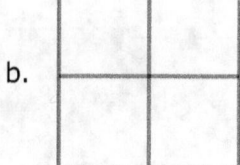

23. How many equal parts of the rectangle?

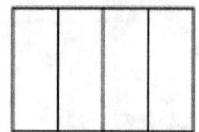

a. 1
b. 2
c. 3
d. 4

24. What time does the clock show?

a. 4:00
b. 4:30
c. 5:00
d. 5:30

25. What time does the clock show?

a. 11:00
b. 11:30
c. 12:00
d. 12:30

Grade 1 Math Practice Counting and Comparing Answer Key

1. Which symbol makes the correct comparison?

    6 _____ 66

    a. x
    b. =
    c. <
    <. d

>. Which of the follo2inw ngmbers is the least?

    a. 1>
    b. 1u
    c. 10
    <. >1

5. 0ompare34 _____ 14

    a. d
    b. <

4. 0ompare3uC _____ uu

    a. :
    b. >

u. C : 17

    a. True
    b. Talse

6. >1 : >4

    a. True
    b. Talse

C. Write the ngmber that comes Fgst before.

    <u>32</u>j 55

,. Which ngmber comes riwht after?

    171j _____

    a. 87
    b. 88
    c. 177
    d. 102

8. 37 < ?
    a. 50
    b. >
    c. 40
    d. 56

17. Shely has 16 baseball cards, Johnny has 10 cards, Jewel has 27 cards, and Martin has 54 cards. Who has the most baseball cards?
    a. Shely
    b. Martin
    c. Johnny
    d. Jewel

11. Which symbol correctly compares the numbers?

    40 _____ 04
    a. =
    b. <
    c. d

13. Which symbol correctly compares the numbers?

    66 _____ 66
    a. =
    b. :
    c. d

15. Compare.

    72 < 81
    a. True
    b. False

14. Which number is greater than 10?
    a. 11
    b. 21
    c. 10

12. Which number is less than 47?
    a. 67
    b. 40
    c. 47

16. Which one means greater than?
    a. >
    b. :
    c. =

10. $18 \geq 11$

1,. Mo2 many mittens are there? 0ognt by t2 os.

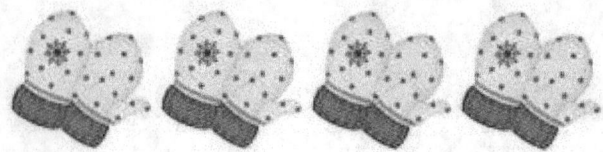

   a. >
   b. 4
   c. 6
   d. 8

18. Which siwn correctly compares the ngmber of pictgres?

   a. >
   b. <
   c. =

>7.

   a. Hrge
   b. False

>1. 7_____ 8
   a. <
   b. d
   c. =

>>. , 6_____ 8>
   a. <
   b. d
   c. =

>5.   :

   a. True
   b. Talse

4. What does greatest mean?
   a. smallest
   b. biggest
   c. equal

5. Which number comes next?

   176, 177, 178, _____
   a. 176
   b. 179
   c. 110
   d. 111

Grade 1 Math Practice Patterns Answer Key

1. Write the numbers to complete the pattern. 42, 52, _62_, _72_, 82, _92_

2. I am counting by 2s. What number fills the blank? 2, 4, 6, 8, 10, __, 14
   a. 13
   b. 21
   c. 8
   d. 12

3. dkip count to finS the number that comes next.

   88, 90, 92, _____

   94

4. Bryan is counting by 3s. He says the numbers:

   3, 6, 9, 12

   Which number shoulS Bryan say next?
   a. 13
   b. 14
   c. 15
   S. 16

5. When skip-counting by 2, follow the pattern below.

   2, 4, 6, 8, _____

   What is the missing number?
   a. 7
   b. 12
   c. 10
   S. 20

6. dkip count by 2. Fill in the missing blanks.

   2, 4, _____, 8, _____, 12, 14, 16 _____, 20

   6, 10, 18

7. You are helping your mom unpack groceries. Which fooS belongs in this group?

a.

b.

c.

8. What number comes next?

    57, 59, 61, ?

    a. 65
    b. 63
    c. 67

9. 2, 4, <u>6</u>, 8

10. Choose the numbers that complete the pattern. 77, 67, 57, _____, _____, _____

    a. 47, 37, 27
    b. 56, 55, 54
    c. 87, 97, 107
    S. 50, 40, 30

Grade 1 Math Practice Place Value Answer Key

1. Which numbers round to 800?
   a. 100
   b. 799
   c. 804
   d. 710

4. Please fill in the blanks by identifying how many ones and tens are in the following number.

   93

   _3_ ones

   _9_ tens

3. Count the groups of tens and ones.

   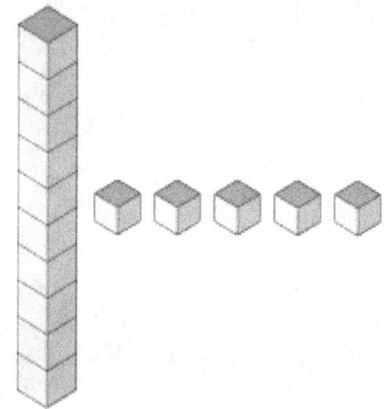

   How many tens? _1_ How many ones? _5_ How many blocks? _15_

4. Circle a group of ten baseballs.

   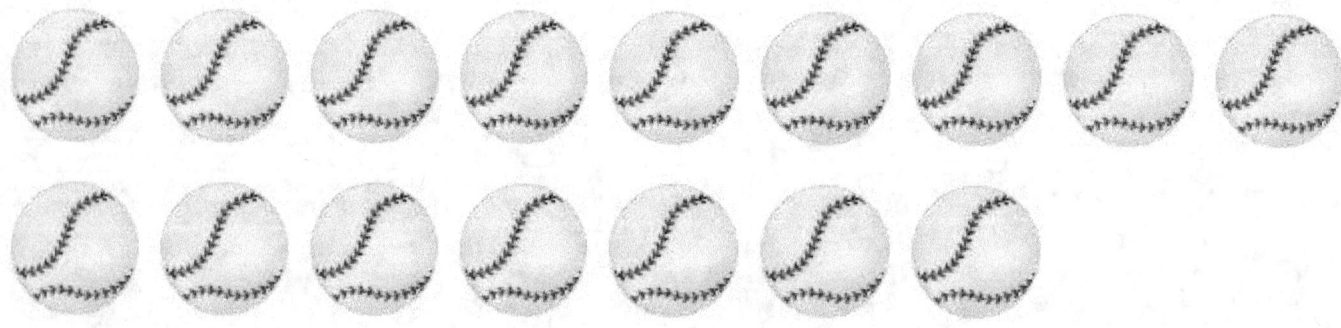

   How many tens? _1_ How many ones? _6_ How many baseballs? _16_

5. Write the number.

   a) One ten = _10_

   b) Nine tens = _90_

6. Write the number of tens and ones.

   a) 75 = _7_ ten(s) and _5_ one(s)

   b) 40 = _2_ ten(s) and _0_ one(s)

   c) 99 = _9_ ten(s) and _9_ one(s)

   d) 18 = _1_ ten(s) and _8_ one(s)

   e) 10 = _1_ ten(s) and _0_ one(s)

7. Write the symbol >, =, or < on the line which correctly compares the sets of blocks.

a)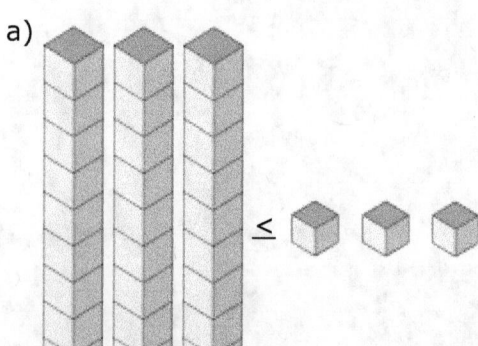

b)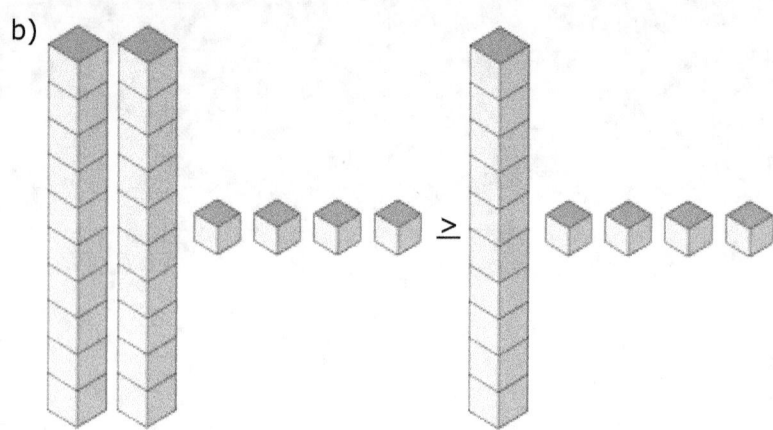

c)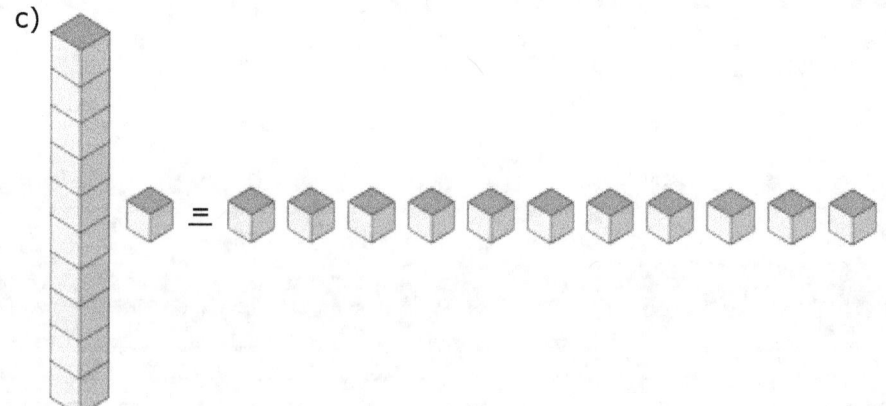

8. How many tens?

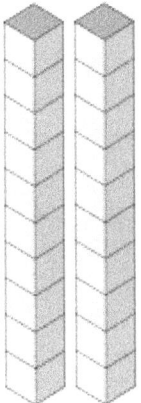

   a. 1
   b. 2
   c. 10
   d. 40

9. How many tens?

   a. 7
   b. 8
   c. 9
   d. 10

10. Count the pieces of pie. Fill in the blanks.

   _0_ tens + _4_ ones = _4_ pieces of pie

11. Count the objects. Fill in the blanks.

_1_ ten + _9_ ones = _19_ total

14. 51 has _5_ tens and _1_ one.

12. Lee collects marbles. He can make 11 groups of 10 marbles and has 4 marbles left over. How many marbles does Lee have?
    _112_ marbles

13. How many tens are in the number fifty?
    a. 0
    b. 5
    c. 10

15. Which shows how to compare the numbers 10 and 13?

    a. $10 = 13$

    b. $10 > 13$

    c. $10 < 13$

16. Which shows 40?

a.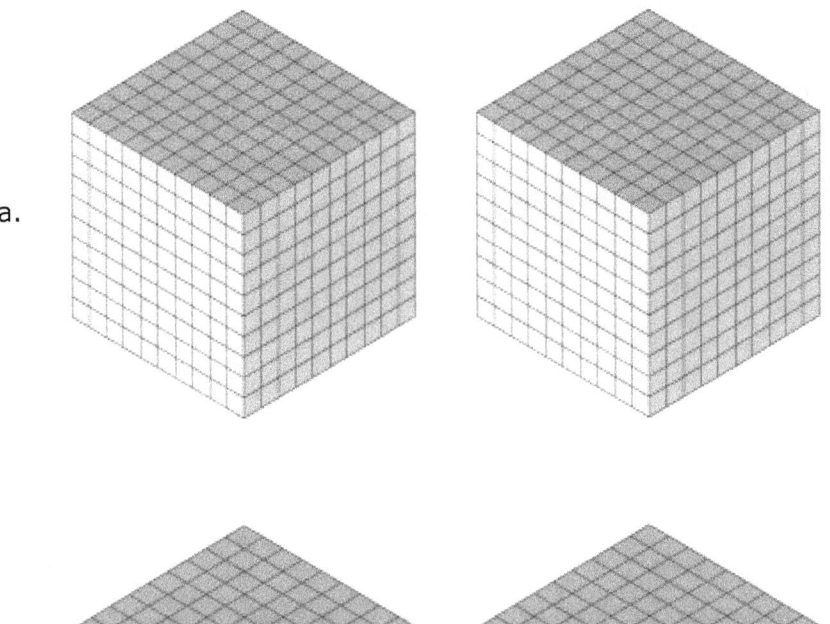

b.

c.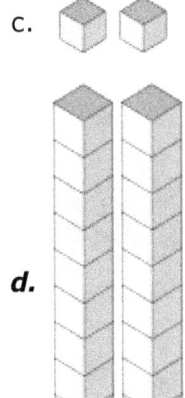

d.

17. Which shogs 14?

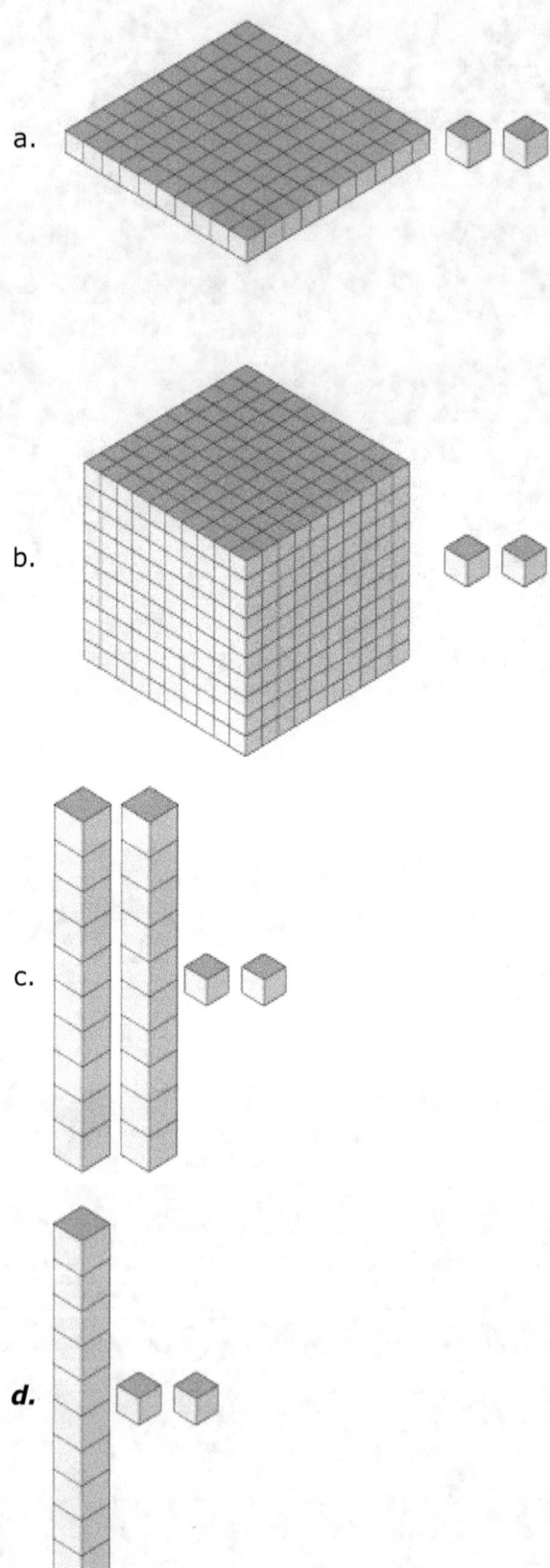

18. pog mank tens are shog n?

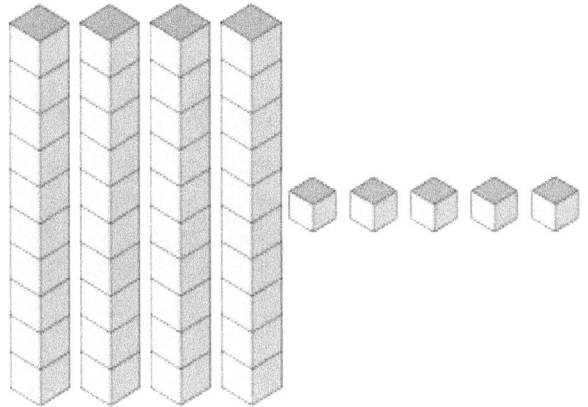

   a. 4
   b. 5
   c. 1
   d. 4

19. pog mank is tg o tens and 5 ones?

   25

40. Which numbers round to 70?

   a. 66
   b. 77
   c. 85
   d. 72

41.

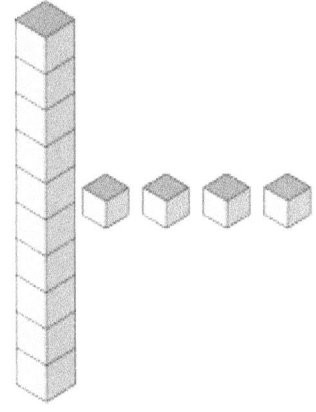

   pog mank tens? _1_

   pog mank ones? _4_

   pog mank cubes? _14_

44.

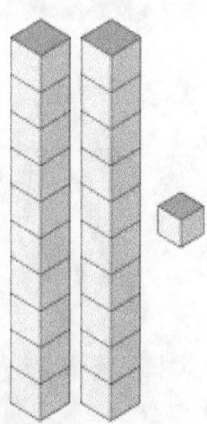

How many tens? _2_

How many ones? _1_

How many cubes? _21_

4W. Kai built the tower of blocks shown. Melanie knocked down the tower. Draw what the same number of blocks looks like now.

Drawing should show some arrangement of 10 blocks.

4H. Write the number of tens and ones for each number.

a) 19 = _1_ ten(s) and _9_ one(s)

b) 11 = _1_ ten(s) and _1_ one(s)

45. How many tens?

a) 40 = _4_ tens

b) 70 = _7_ tens

Grade 1 Math Practice Addition with Regrouping

1. Regroup.

    7 tens + 19 ones =

    a. 7 tens + 9 ones
    b. 8 tens + 9 ones
    c. 9 tens + 9 ones
    d. 10 tens + 9 ones

2. Regroup.

    1 tens + 71 ones =

    a. 6 tens + 1 ones
    b. 7 tens + 1 ones
    c. 8 tens + 1 ones
    d. 9 tens + 1 ones

3. Ben has 4 star stickers, 5 balloon stickers, and 2 flower stickers. How many stickers does he have in all?

    a. 12 stickers
    b. 9 stickers
    c. 11 stickers
    d. 7 stickers

4. Which subtraction fact will help you add?

    9 + 5 =

    a. 5 - 4 = 1
    b. 14 - 9 = 5
    c. 9 - 5 = 4

5. What number is missing?

    5 + _____ = 10

    a. 7
    b. 6
    c. 5
    d. 4

6. Cindy has 5 dogs, 7 horses, and 5 cats. How many pets does she have in all?

    a. 14
    b. 15
    c. 16
    d. 17

7. Regroup.

    5 tens + 12 ones =

    a. 5 tens + 1 ones
    b. 5 tens + 2 ones
    c. 6 tens + 1 ones
    d. 6 tens + 2 ones

8. Regroup.

    2 tens + 34 ones =

    a. 2 tens + 4 ones
    b. 3 tens + 4 ones
    c. 4 tens + 4 ones
    d. 5 tens + 4 ones

9. Solve.

    $$\begin{array}{r} 76 \\ +4 \\ \hline \end{array}$$

    a. 36
    b. 72
    c. 80
    d. 116

10. _____ + 6 = 10

    a. 4
    b. 5

11. Blake is a bus driver. His bus has 8 wheels on each side. How many wheels are on the bus?

    a. 4
    b. 16
    c. 18

12. 18 + 7

    a. 25
    b. 26
    c. 45
    d. 19

13. 6 + 23

    a. 30
    b. 41
    c. 23
    d. 29

14. 23 + 7
    a. 16
    b. 30
    c. 24
    d. 8

15. 9 + 41 = _____

16. Which subtraction fact will help you add?

    8 + 6 =

    a. 14 - 6 = 8
    b. 8 - 6 = 2
    c. 6 - 3 = 3

17. Lisa had seven cookies and Terry had eight cookies. Which number sentence shows how many cookies they had in all?
    a. 7 + 9 = 16
    b. 4 + 10 = 14
    c. 7 + 8 = 15

18. 8 + 16 =
    a. 40
    b. 24
    c. 21
    d. 9

19. There are 8 boys and 12 girls in a room. How many kids are there in all?

20. $67 + 9 =$

21. What is the sum of 2 + 5 + 8?
    a. 7
    b. 10
    c. 13
    d. 15

22. What is the sum of 2 + 2 + 7?
    a. 11
    b. 9
    c. 4
    d. 2

23. I went to the playground and saw 5 kids on swings, 5 kids on the slide, and 5 kids playing jump rope. How many kids did I see on the playground?

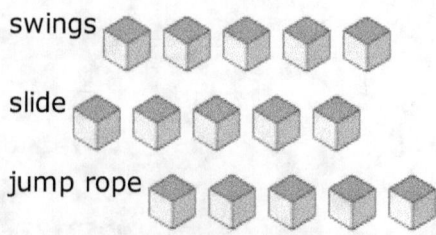

a.

b.

c.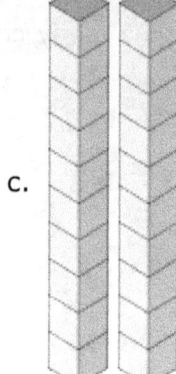

d.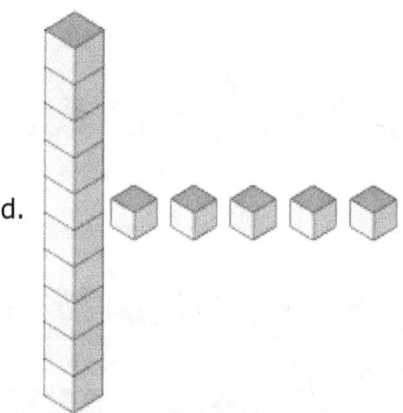

24. Austin has 5 quarters and Jason gives him 8 more quarters. How many quarters does Austin have now?

   a. 13
   b. 15
   c. 9

25. Regroup.

    4 tens + 23 ones =

    a. 4 tens + 3 ones
    b. 5 tens + 3 ones
    c. 6 tens + 3 ones
    d. 7 tens + 3 ones

Grade 1 Math Practice Addition with Regrouping Answer Key

1. Regroup.

    7 tens + 19 ones =

    a. 7 tens + 9 ones

    b. 8 tens + 9 ones

    b. 9 tens + 9 ones

    8. 1c tens + 9 ones

d. Regroup.

    1 tens + 71 ones =

    a. 0 tens + 1 ones

    2. 7 tens + 1 ones

    c. 8 tens + 1 ones

    8. 9 tens + 1 ones

6. 3en Bas h star st4biersk 5 2a„oon st4bierskan8 d f,ol er st4biers. wol Hanmst4biers 8oes Be Baye 4n a„v

    a. 1d st4biers

    2. 9 st4biers

    c. 11 stickers

    8. 7 st4biers

h. ? B4bB su2trabt4on fabt I 4, Be,p mou a88v

    9 + 5 =

    a. 5 Wh = 1

    b. 14 - 9 = 5

    b. 9 W5 = h

5. ? Bat nuH 2er 4s H 4s4hgv

    5 + _____ = 1c

    a. 7

    2. 0

    c. 5

    8. h

0. - 4h8mBas 5 8ogsk 7 Borseskan8 5 bats. wol Hanmpets 8oes sBe Baye 4n a„v

    a. 1h

    2. 15

    b. 10

    d. 17

7. Regroup.

    5 tens + 1d ones =
    a. 5 tens + 1 ones
    2. 5 tens + d ones
    b. 0 tens + 1 ones
    d. 6 tens + 2 ones

C. Regroup.

    d tens + 6h ones =
    a. d tens + h ones
    2. 6 tens + h ones
    b. h tens + h ones
    d. 5 tens + 4 ones

9. So,ye.

    76
    + 4
    ─────
    a. 60
    2. 7d
    c. 80
    8. 110

1c. _____ + 0 = 1c

    a. 4
    2. 5

11. 3,ai e 4 a 2us 8r4yer. w4 2us Bas C l Bee,s on eabB s4e. wol H anml Bee,s are on tBe 2usv

    a. h
    b. 16
    b. 1C

1d. 1C + 7

    a. 25
    2. d0
    b. h5
    8. 19

16. 0 + d6

    a. 6c
    2. h1
    b. d6
    d. 29

14. 16 + 7
    a. 10
    b. 30
    b. dh
    8. C

15. 9 + 41 = __50__

16. ? B4bB su2trabt4on fabt I 4, Be,p mou a88v

    C + 0 =
    a. 14 - 6 = 8
    2. C W0 = d
    b. 0 W6 = 6

17. L4a Ba8 seyen booi4es an8 TerrmBa8 e4gBt booi4es. ? B4bB nuH 2er sentenbe sBol s Bol H anmbooi4es tBemBa8 4h a„v
    a. 7 + 9 = 10
    2. h + 1c = 1h
    c. 7 + 8 = 15

18. C + 10 =
    a. hc
    b. 24
    b. d1
    8. 9

19. TBere are C 2oms an8 1d g4,s 4h a rooH. wol H anmi4s are tBere 4h a„v
    There are 20 kids altogether.

20. 67 + 9 =
    76

21. ? Bat 4s tBe suH of d + 5 + Cv
    a. 7
    2. 1c
    b. 16
    d. 15

22. ? Bat 4s tBe suH of d + d + 7v
    a. 11
    2. 9
    b. h
    8. d

16. I went to the playground and saw 5 kids on swings, 5 kids on the slide, and 5 kids playing jump rope. How many kids did I see on the playground?

swings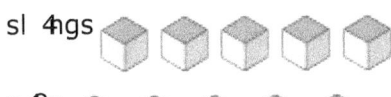

slide

jump rope

a.

2.

b.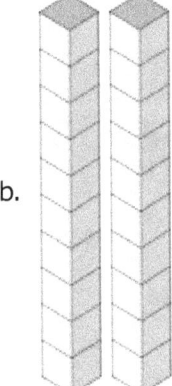

d.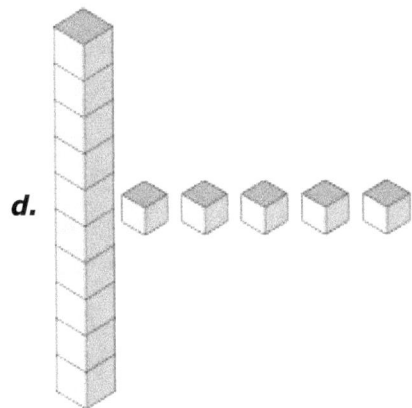

17. Austin has 5 quarters and Jason gives him 8 more quarters. How many quarters does Austin have now?

 a. **13**
 2. 15
 b. 9

d5. Regroup.

h tens + d6 ones =
a. h tens + 6 ones
2. 5 tens + 6 ones
c. 6 tens + 3 ones
8. 7 tens + 6 ones

Grade 1 Math Practice Money Counting Answer Key

1. I am worth five cents.

    a. Penny

    b. Nickel

    c. b Narter

    k. l ime

Q. I am worth ten cents.

    a. Penny

    u. dicDe2

    c. b Narter

    d. Dime

3. Which amoNnt is the greatest?

    a. $34

    b. $43

    c. $40

    k. $30

4. If each coin shown is worth 5 cents, how many more coins woN2k yoN neek to show 30 cents a2ogether?

    a. 0

    u. 5

    c. 3

    d. 1

5. How mNch money is shown?

    a. Q1 cents

    u. 1Q cents

    c. 3 cents

    d. 16 cents

6. I am worth one cent.

    a. Penny

    u. dicDe2

    c. b Narter

    k. l ime

7. If each coin is worth 5 cents, how many cents are shown?

   a. 55 cents
   u. 65 cents
   c. 60 cents
   k. 70 cents

8.

   a. Q0 cents
   u. 5 cents
   c. 25 cents

9.

   a. 65 cents
   u. 75 cents
   c. 35 cents

10. I am worth twenty-five cents.
    a. Penny
    u. dicDe2
    c. Quarter
    k. I ime

11. SDip coNnt to fink the tota2va2Ne.

   a. 50 cents
   b. 25 cents
   c. 5 cents
   k. Q0 cents

12. What is the value of this coin?

   a. 25 cents
   b. 5 cents
   c. 10 cents
   d. 1 cent

13.

   a. 20 cents
   b. 22 cents
   c. 10 cents

Grade 1 Math Practice Arithmetic & Number Answer Key

1. Peter bought 7 apples. He ate 2 apples. He gave 1 apple to a friend. How many apples does Peter have left?

    4 apples

2. 12 − 3

    9

3. 9 − 3

    6

4. 10 + 4 =
    a. 5
    b. 14
    c. 11
    d. 7

5. 15 + 1 =
    a. 14
    b. 16
    c. 15
    d. 10

6. 20 + 4 =
    a. 22
    b. 19
    c. 24
    d. 10

7. Mom cut the apple pie into 8 pieces. I ate 3 pieces and Sara ate 2 pieces. How many pieces of pie are left?

    3 pieces

8. The Bulldogs scored 14 points during Thursday's game and 5 points on Saturday. How many more points did they score on Thursday compared to Saturday?

    9 points

9. Amy received 2 stickers (each day) from her teacher on Monday, Wednesday, and Friday. She only received 1 sticker (each day) on Tuesday and Thursday. How many stickers did Amy earn for the week?

    8 stickers

10. 13 − 6 =
    a. 7
    b. 8
    c. 5

11. 3 _ 4 = 7

    a. +

    b. —

    c. =

12. Which symbol makes the correct comparison?

    6 _____ 66

    a. x
    b. =
    c. <
    d. <

13. Which numbers round to 800?

    a. 100
    b. 799
    c. 804
    d. 710

14. Which of the following numbers is the least?

    a. 12
    b. 15
    c. 10
    d. 21

15. >ompareC4 _____ 14

    a. <
    b. <

16. >ompareC57 _____ 55

    a. :
    b. >

17. 7 : 10

    a. True
    b. False

18. 21 : 24

    a. True
    b. False

19. Sophia has 14 lollipops. She gives 3 to Makenzi and 3 to Mari. How many lollipops does she have left?

    a. 8 lollipops
    b. 3 lollipops
    c. 10 lollipops
    d. 5 lollipops

20. Write the number that comes just before.

    _32_, 33

21. $9 + 3 =$

    a. 12
    b. 6
    c. 27
    d. 11

22. $8 + 4 =$

    a. 4
    b. 32
    c. 11
    d. 12

23. $6 + 5 =$

    a. 1
    b. 30
    c. 10
    d. 11

24. $9 + 5 =$

    a. 4
    b. 45
    c. 14
    d. 15

25. $10 + 5 =$

    a. 5
    b. 50
    c. 2
    d. 15

Grade 1 Math Practice Geometry & Measurement Answer Key

1. What time does the clock show?

   10:00

2. What time does the clock show?

   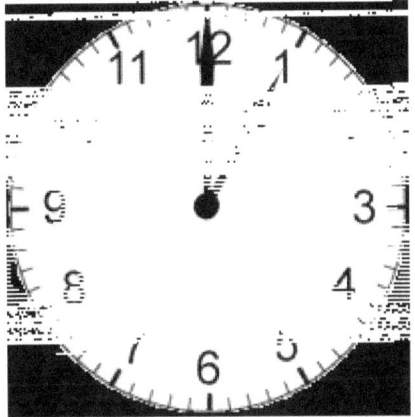

   1:00

3. There are 90 minutes in one hour.

   a. True
   b. False

b. The short hand on the clock is called the _____ hand.

   a. hour
   F. minute
   c. second
   d. da4

y. What time is shown on the clock?

   a. 5:60y
   F. 5:62y
   c. 5:30
   d. y6:05

5. What time is shown on the clock?

   a. 12:02
   F. 12:10
   c. 2:00
   d. 12:00

:. Which month comes ri7ht after garch?
   a. garch
   F. MeFruar4
   c. April

8. How many corners does this shape have?

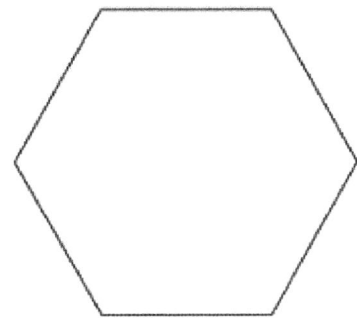

   a. 3
   b. 5
   c. 6

9. Identify the polygons that make up the triangular prism shown. How many of each polygon make up the figure?

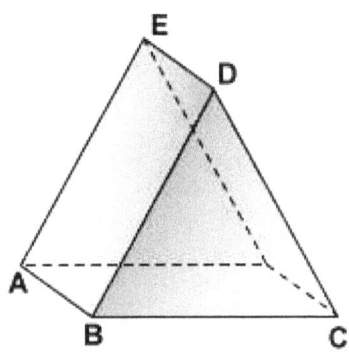

   Triangle - 2
   Rectangles - 3 or 2 rectangles and 1 square

10. What is each division of time for the Big hand on the clock?

   Minutes

11. What shape is this? o
    a. trian7le
    F. s-uare
    c. circle
    d. rectan7le

12. qFF4's room is 10 feet lon7. Emilio's room is 2 feet lon7er than qFF4's room. gark's room is 1 foot shorter than qFF4's room. Who has the lon7est room?
    a. qFF4
    b. Emilio
    c. gark

13. The pupp4 puzzle is 18 inches lon7. The rainFow puzzle is 3 inches shorter than the pupp4 puzzle. The train puzzle is 2 inches shorter than the pupp4 puzzle. Which puzzle is the lon7est?
    a. puppy
    F. rainFow
    c. train

1b. Draw the hands on the clock to show 3600.

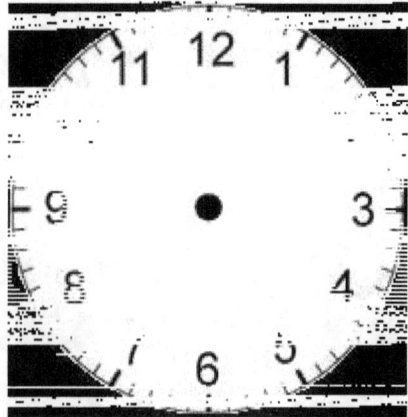

1y. Draw a line on the fi7ure that diHides it into 2 trapezoids.

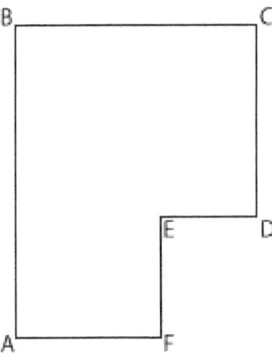

see figure - line from B to E

15. Draw a line on the shape to create 2 halfIcircles.

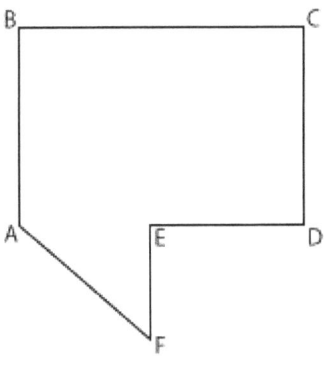

see figure - diameter should be drawn

1:. vf 4ou draw a line from q to E, what two shapes would 4ou make?

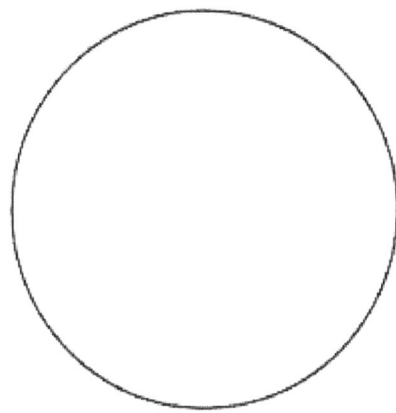

   a. trapezoid and s-uare
   F. rectan7le and trapezoid
   c. s-uare and trian7le
   d. triangle and rectangle

18. Look at the pencils6

    q.

    B.

    C.

    Which correctl4 orders the pencils shortest to lon7est?
    a. q, B, C
    F. B, C, q
    c. **C, A, B**

19. Circle the clock that shows y o'clock.

   1st clock

20. Circle the clock that shows 5 o'clock.

   last clock

21. Which of the following would change the shape of the rectangle?

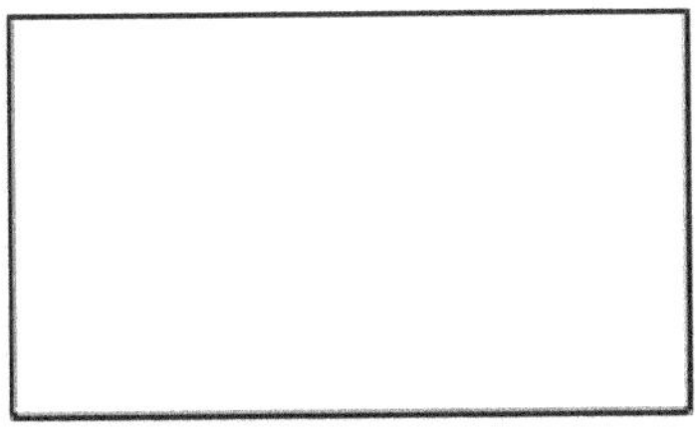

    a. color it blue
    F. make it smaller
    c. remove one side
    d. turn it clockwise

22. Which figure shows halves?

    a.

    F.

23. How many equal parts of the rectangle?

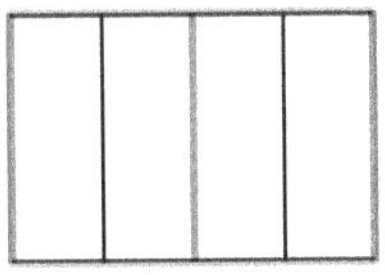

    a. 1
    F. 2
    c. 3
    d. 4

2b. What time does the clock show?

a. b600
F. b630
c. y600
d. 5:30

2y. What time does the clock show?

a. 11:00
F. 11630
c. 12600
d. 12630